Crossfuze
Enterprise Services That Win

AF824726

10 Pillars of Customer Service Management Success

Essential Strategies for Modernizing Customer Service and Accelerating Bottom-Line Profitability

This book is for informational purposes only and does not constitute legal advice. If you require legal advice, retain a lawyer licensed in your jurisdiction. The opinions expressed are solely those of the authors, who are not attorneys.

 Published by Crossfuze.

For information, address:

Crossfuze, 500 Seneca St. #620, Buffalo, NY 14204.

ISBN: 978-0-9993184-0-9

Visit www.crossfuze.com

Crossfuze
Enterprise Services That Win

Table of Contents

Introduction

If your customer service operation is like most that are stuck at the turn of the last century, your customer experience (CX) capabilities are limited at best—and completely archaic at worst. The challenge is that customer service has for decades relied on resolving customer inquiries one at a time, and then painstakingly logging the cases into multiple disconnected systems.

Thanks to the Customer Service Management (CSM) application from ServiceNow, that outdated paradigm has been turned upside down: Contact center teams are now able to effectively manage the CX through the full customer life cycle, creating strategic interconnectivity and dynamic engagement around all customer workflows and processes. ServiceNow CSM provides CX leaders access to powerful capabilities that can efficiently analyze the root causes of customer problems while delivering valuable insights based on aggregating disparate pieces of customer data.

What this means for the bottom line is an unprecedented opportunity to re-engage customers and to deliver personalized experiences that exceed customer expectations, rapidly getting your customers to appropriate resolutions. In an era of social media and online reviews, delivering these customer service experiences with consistency across the full customer life cycle is more important than ever. CX plays a disproportionate role in shaping how your company is perceived and, in turn, how you focus your branding and marketing efforts.

While ServiceNow CSM has the technical power to reimagine contact centers in just about any manner, you need vision and strategy to transform how your contact center operates and engages with customers.

> The purpose of this book, the *10 Pillars of Customer Service Management Success*, is to ensure essential thought processes, goals, and priorities are kept front and center during your CX transformation journey. You will learn the 10 most crucial elements, or pillars, for modernizing customer service and driving bottom-line profitability for your company.

The *10 Pillars of Customer Service Management Success* is a distillation of insights, lessons learned, and working knowledge from Crossfuze experts who have run contact centers for over 25 years, as well as helped many other organizations transform their contact centers. The perspectives and guidance offered by Crossfuze experts are based on both rich operational experiences and the technical know-how of optimally configuring and customizing the ServiceNow CSM application. Our CSM experts focus on sharing proven practices that will help you achieve rapid CSM victories and make optimal use of your limited resources and bandwidth.

Admittedly, a single book can't replace years of hard-earned experience or offer in-depth blueprints on how to reimagine your contact center operations and customer experiences with your brand. Therefore, if at any time you would like to do a deep dive into your specific challenges and objectives for transforming your contact center, the Crossfuze team is here to listen and help when and where needed.

Pillar 1

Establishing Your Contact Center's Value

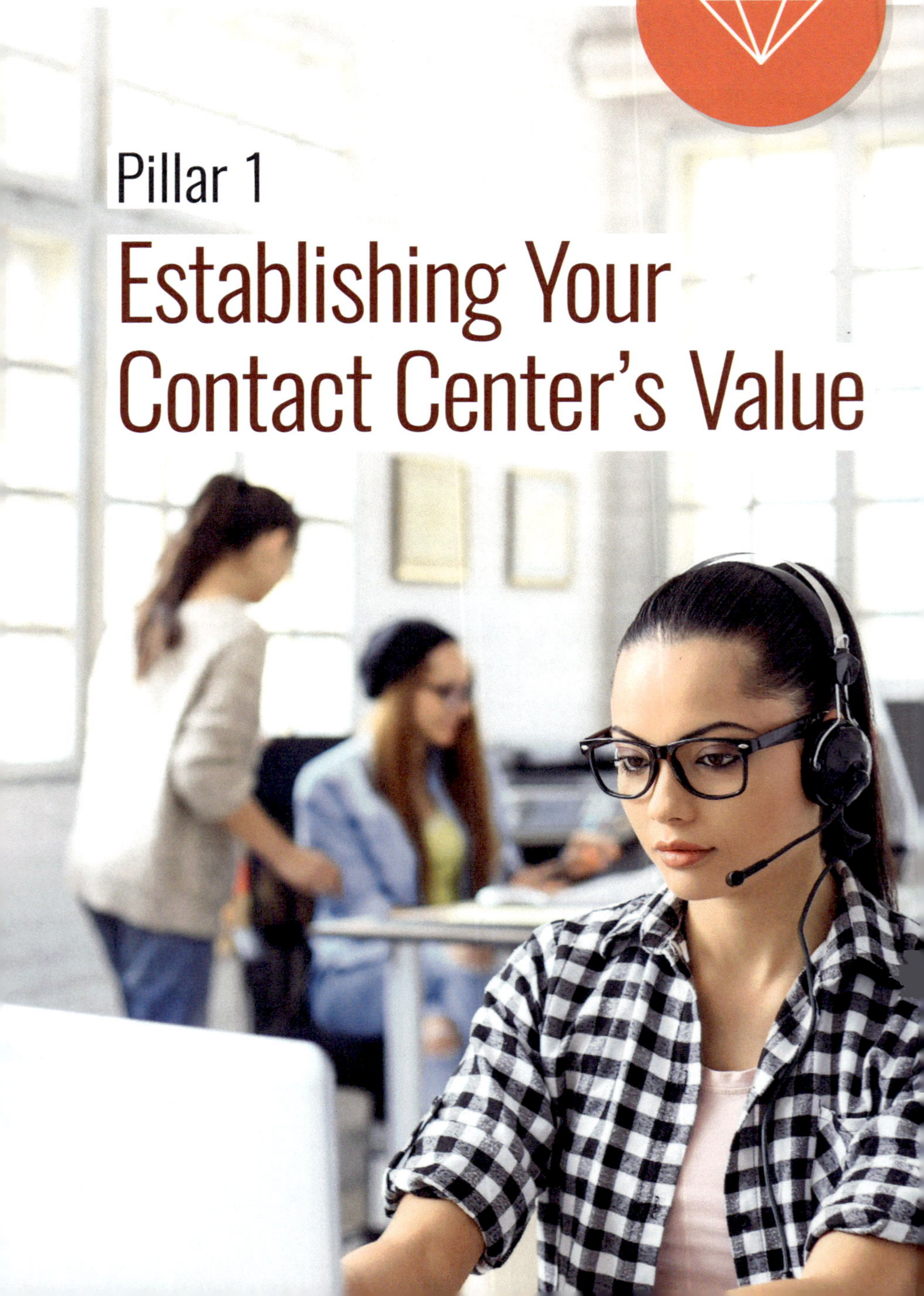

MEET TONY

Tony is a longtime vice president of customer service whose contact center is the target of constant skepticism and financial review by the C-suite. He recognizes that such attitudes toward contact centers are endemic across all industries; indeed, many organizations think of their contact centers as necessary overhead expenses. Tony understands, however, that modernized contact centers are essential in driving brand loyalty and additional sales. In fact, Tony already has a vision for modernizing his contact center using the Customer Service Management (CSM) platform from ServiceNow. How does Tony build support for his plans and change deeply ingrained attitudes about the business value of his contact center?

The solution to Tony's dilemma is to establish a new positioning for the contact center across the enterprise. This change of thinking is essential for getting a green light from the C-suite and gaining buy-in from internal end users. Unfortunately, Tony's struggle to build support for improving his contact center is not unusual. Many organizations still view their contact centers as an inescapable overhead cost, and consequently, undervalue and marginalize this key business asset. A 2015 study by the International Customer Management Institute found that 71 percent of contact center representatives say that upper management doesn't provide them with the systems and tools they need to do their jobs effectively.

Customer service leaders want to take a number of proactive steps to overcome misperceptions about contact centers, and ultimately to convince others of the strategic value of contact centers. This legwork is foundational to moving forward with any CX modernization effort, especially on an accelerated timeline. Having a clear vision for CX motivates organizations to take full advantage of ServiceNow CSM's powerful capabilities and create an essential competitive advantage. Let's explore five essential strategies for establishing the value of your contact center across the enterprise.

71% of contact center representatives say that upper management doesn't provide them with the systems and tools they need to do their jobs effectively

1.1 OPTIMIZE STAFFING FOR YOUR CONTACT CENTER:

The first step to establishing a contact center's value is to ensure existing resources are optimally allocated. If your own contact center staff can articulate deficiencies in quality and speed of customer service delivery, it's crucial that you pinpoint what those deficiencies are and work toward solutions. In some cases, the staffing optimization process may be more nuanced. For example, some representatives will be better at resolving certain types of customer inquiries than others. You want to strategically shuffle your staffing so that you can deliver a strong CX to as many individual customer cases as possible, especially your high-value customers whom you will be supporting and engaging with over the long term.

1.2 CONDUCT ANALYTICS ON YOUR CONTACT CENTER DATA TO GAIN INSIGHTS:

Your contact center has amassed a tremendous amount of customer data. You can use analytics to aggregate that data and identify important trends, opportunities, and potential problems. Analytics gives you meaningful insights that you can then use to drive improvements to customer service operations, and to offer more of the features that customers want. When you're able to demonstrate that you can get actionable insight from contact center data, it changes employees' perceptions of the overall value of the contact center, especially when improvements uncover new sales streams.

1.3 SHARE THE CONTACT CENTER'S WEALTH OF CUSTOMER DATA:

Customer data generated by a contact center is useful to more than just the contact center itself. Contact centers generate invaluable data on customer spending habits, customer preferences and priorities, product and service quality and value, and basic demographic information. Other departments in the organization can use this data to, for example, develop more effective marketing campaigns, promote new product/service features, build strong relationships with customers over the full customer life cycle, and engage in long-term strategic planning and goal-setting.

1.4 GIVE YOUR CONTACT CENTER ACCESS TO CORPORATE SOCIAL MEDIA ACCOUNTS:

If your social media channels are run by your marketing or communications teams, you want to rethink this division of responsibilities. While these folks may think they know how to engage your customers, your contact center spends all day, every day, engaging with customers one on one. Your contact center team knows exactly how customers feel, what rubs them the wrong way, and what messages and information they most want to hear from the business at any given time. Your contact center can use these "pulse on the customer" insights to share information, answer customer questions, and engage with both loyal customers and critics. Your contact center also can work alongside your marketing and communications teams to improve and refine their social media engagement efforts.

1.5 EDUCATE THE C-SUITE ON CONTACT CENTER KEY PERFORMANCE INDICATORS (KPIs):

Contact centers aren't evaluated solely on their customer satisfaction scores. There are numerous KPIs that offer key insights into the value that contact centers provide to the organization as a whole, and these insights should be highlighted constantly to upper management. Examples of contact center KPIs that matter to the C-suite include first call resolution rate, cost per transaction, transactional value, and adherence to budgets. Because the C-suite thinks in KPIs, it's strategically important to explain the value of a contact center in these terms. Some of these metrics establish the value proposition of call deflection, a key feature of the CSM application that tends to particularly resonate with the C-suite.

Contact centers offer so much more than just a way to support customers who have questions and issues. Contact centers deliver invaluable knowledge to every facet of the company, and have the most direct and sustained daily relationship with customers. To help illuminate this value to others in your organization, especially the C-suite, you want to optimize staffing assignments within your contact center, conduct analytics on contact center data, share contact center data widely and often, get contact center agents involved in daily moderating and engaging with customers on corporate social media channels, and discuss the contact center's value with the C-suite in terms of overall objectives and KPIs.

How confident are you in answering these questions?

- How can your contact center demonstrate its strategic business value?
- What customer data is your contact center prepared to share with others across the enterprise?
- What inefficiencies can you identify in your contact center staffing resources?

Pillar 2

Forging a Clear Path to Achieve Your CSM Goals

MEET COLLIN

Collin is a newly minted head of customer service with an enormous responsibility on his shoulders: He's in charge of modernizing his company's customer service operation through the use of technology. The C-suite recognizes that customer service has fallen behind the technology curve, creating a competitive disadvantage that Collin is expected to immediately reverse. At the same time, the C-suite expects Collin to lead this ambitious undertaking in a strategic, cost-effective fashion. Collin feels compelled to take action, but he's not sure his approach will plot an optimal path forward and align with corporate strategy.

ABOUT HALF of all businesses report that they've experienced an IT project failure within the past year.

For Collin and countless other leaders preparing to modernize customer service operations, the stakes are high. About half of all businesses report that they've experienced an IT project failure within the past year, and overall, the average portion of all IT projects that are deemed failures sits at 14 percent, according to a 2016 Project Management Institute survey. This means that on average, one in every seven IT projects is deemed a failure. To ensure you put your CSM implementation on a path to success, there are several essential actions you want to take to guard against costly failure. These best-practices solutions have been time-tested and proven effective by the many organizations that have come before yours. In fact, credible solutions from credible sources are essential to the long-term success of CSM. Let's explore five essential tips for forging a clear path forward on your journey to a CSM transformation.

2.1 ARTICULATE A VISION OF WHAT SUCCESS LOOKS LIKE:

To maximize the effectiveness of any enterprise initiative, it's crucial that you develop a well-defined, clear vision of what you're trying to achieve. The first question you should ask yourself is: How can we modernize CSM to transform CX and drive bottom-line profitability? The answer to this question is your big-picture outcome. From there, you can work backward and define how you'll maximize use of your resources and talents to achieve this vision. When you have a clear vision, you're able to build support for what you're trying to achieve, because others can see the big-picture outcomes and become excited by them.

2.2 BUILD A BUSINESS CASE THAT TIES CSM TO STRATEGIC BUSINESS GOALS:

You are not modernizing CX in a vacuum; all of the bells and whistles you implement should be linked to the organization's long-term business strategy. In other words, every element of CSM should advance the organization's master plans in some way. When you articulate this relationship, you're building a business case that establishes CSM as a priority. Your business case, not coincidentally, becomes a message that resonates with the C-suite, which is constantly evaluating all projects based on their strategic business value.

2.3 BUILD AN AMBITIOUS BUT MALLEABLE IMPLEMENTATION ROADMAP:

Your CSM implementation journey should not be something you make up on the fly. Before you embark on your journey, you want to flesh out a detailed implementation roadmap that articulates all of the specific steps and milestones you aim to hit over the next several years. Not only will creating this roadmap help you focus and improve your own planning processes, but it also will help others to understand what to expect at various points and how individual tasks fit into the big picture. Finally, be sure to design your roadmap with full intention of changing it later; your roadmap needs to be flexible and adaptable as unanticipated issues arise.

2.4 GUARANTEE EARLY WINS TO MAINTAIN MOMENTUM:

As you develop your CSM vision and implementation roadmap, you want to stack the deck in your favor by building early, guaranteed victories into the journey. Early victories are the incremental home runs that will motivate and inspire your team to continue working toward a CX transformation. By deliberately planting these guaranteed wins on your path, you are buffering against unanticipated setbacks and problems that can derail momentum and enthusiasm. Early wins also serve as validation to you and others, including your skeptics and naysayers, that the implementation journey you've selected is the right one for your enterprise.

2.5 EVANGELIZE YOUR PLANS FAR AND WIDE:

Your success with CSM is directly dependent on your ability to get buy-in for your vision and roadmap. To maximize the support you receive, you want to talk up your plans with as many influencers in your organization as possible. When they're excited and invested, they instinctively amplify your messages far and wide, enabling you to reach far more people than you could do alone. You can evangelize by hosting a kickoff meeting with key influencers that outlines strategic thinking and goals behind your decision. You can follow up this meeting with check-in points at routine intervals to keep the excitement high and to ensure goals and objectives remain clear in the minds of all your influencers. Your goal is to continuously report on your victories and breathe life into your CSM journey—even two or three years down the road.

Beginning your CSM implementation journey can seem daunting, but remember that expert solutions are available from the successes that have preceded your CSM journey. You want to tap into that wealth of knowledge by turning to trusted resources and experienced consultants whose expert guidance will keep you on track and focused. The journey to CSM transformation starts with articulating a vision of what success looks like, building a business case that ties CSM to strategic business goals, building an optimal implementation roadmap, setting up your team for early wins, and evangelizing your plans to build support and buy-in.

How confident are you in answering these questions?

- What does a CSM transformation look like for your organization?
- What is your organization's bottom-line business case for CSM?
- Which potential CSM wins do you believe could be guaranteed home runs for your organization?

Early victories are the incremental home runs that will motivate and inspire your team to continue working toward a CX transformation. By deliberately planting these guaranteed wins on your path, you are buffering against unanticipated setbacks and problems that can derail momentum and enthusiasm.

Pillar 3

Understanding What Your Customers Want

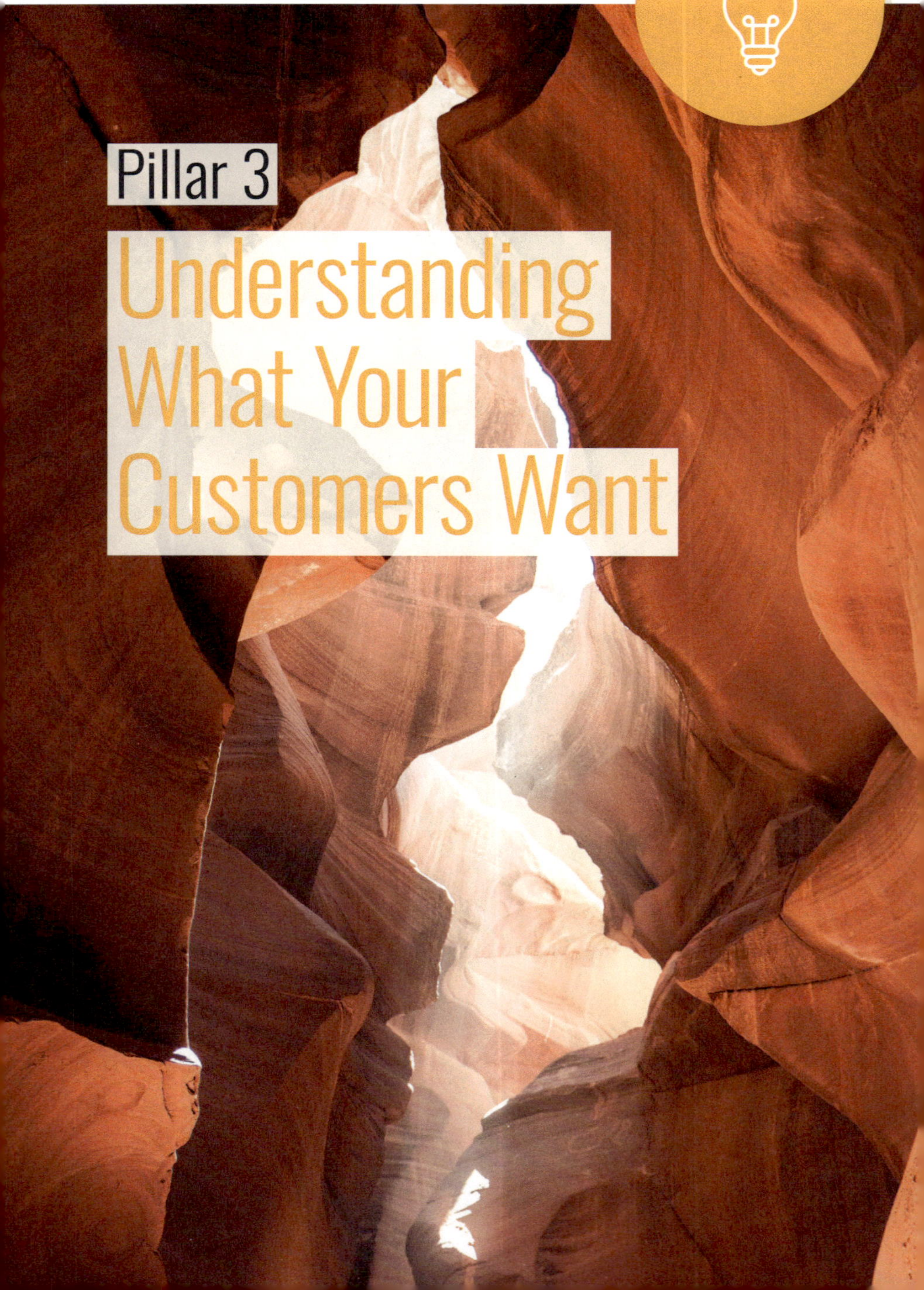

MEET TIM

Tim works as vice president of customer experience for a Fortune 500 company. He's traditionally focused on managing a sprawling, decentralized network of contact centers that span multiple continents. Tim has achieved some success reducing average customer wait times and boosting CSAT scores, but customer service remains largely a reactive, ticket-oriented experience across the enterprise: Customers call in, a case file is created or updated, and the agent works as quickly as possible to resolve the issue or escalate to the next tier. Tim knows that CX has evolved markedly over the past decade, and that ServiceNow's leading Customer Service Management (CSM) solution enables companies to maintain vast treasure troves of data on their customers and to build meaningful, personalized relationships with customers over the full customer life cycle. As Tim explores the CSM platform, he feels overwhelmed and unsure of how to use it to the company's advantage. How should he reimagine the delivery of customer service? What are his company's customers looking for and expecting when they initiate contact? How can CX move customers from satisfied to delighted—and keep them that way over the long term?

The trepidation that Tim feels is common when first embarking on the journey to implement ServiceNow CSM. In fact, the stakes are quite high: Less than one in five U.S. customers report that their customer experience was good or excellent, according to a 2016 survey from the research firm Forrester. Conversely, 59 percent of customers rate their CX just "OK," and 23 percent rate their CX as poor or very poor. Businesses, however, are increasingly recognizing the importance of CX in driving customer loyalty and profitability. In the 2016 Forrester survey, 71 percent of businesses reported that they are prioritizing CX improvements.

Over the years, the technology that powers customer service has evolved dramatically, and the expectations and demands of modern-day customers have evolved right along with the technology. Today, the most effective strategy for driving improvements to CX is to recognize what your customers want and expect when they initiate contact, and then to align your CSM solution to your customers' values and priorities. Let's explore the six most important things you should focus on to align your CSM solution with your customers' values and priorities.

3.1 CREATE AN OMNICHANNEL CONTACT CENTER:

When customers encounter problems or need help, they don't all intuitively pick up the phone and call. The vast majority now turn to email, look for a live chat option, or reach out over social media. Some don't want to engage with a live representative at all: They seek out online self-help resources for the answers to their questions. And commonly, customers will use a combination of channels to seek out desired information, depending on urgency. This trend holds true even for customers who already have an established relationship with the company. Therefore, it's important to offer multiple support channels and ensure these channels are integrated into one data management system. By doing so, you're able to have a holistic view of the customer's support request activities, which helps to provide the customer with a consistent and seamless CX, regardless of what channel(s) they use.

3.2 FOCUS ON DELIVERING THE FASTEST RESOLUTIONS POSSIBLE:

No one wants to post a question on a contact center channel only to get a response days later or, worse yet, have no one respond at all. Likewise, customers never are keen on being put on hold, transferred from one department to another, or stuck with a misinformed, ill-prepared representative. Not only is it common sense to deliver the fastest resolutions possible no matter what channel of communication they use, but it's also mandatory. We live in an era where customers are accustomed to instant gratification. Customer service that's perceived as falling short of maximum efficiency will be swiftly judged. If that judgment is not in your favor, you risk losing that customer to a competitor. One critical way to maintain focus on response times is by implementing Service Level Agreements (SLAs) that define targets for how quickly you will respond. SLAs can then serve as your standardized metrics against which to judge success.

3.3 OFFER INTUITIVE SELF-SERVICE OPTIONS:

Most customers who have grown up with the Internet genuinely believe they should be able to Google the answer to any question they have. A recent survey by the firm Research Now found that 59 percent of customers become frustrated when they must reach out to a live agent because they cannot find a resolution to their issue online. You want to rise to this challenge by creating self-service customer service portals that contain answers to commonly asked questions; in fact, as you build your self-service portal using ServiceNow, you want to design

it around precisely the issues that your customers are most likely to search for online. Because self-service can dramatically reduce CX costs, self-service is more than just good for the customer; it's also good for your bottom line.

3.4 USE THE RESOURCES OF THE FULL ENTERPRISE TO DELIVER CUSTOMER SOLUTIONS:

When customers make contact with a business, they expect that their initial point of contact will do more than take a message and/or pass them through to another representative. Modern-day customers expect that their initial point of contact has access to the full resources and expertise of the entire organization—and all from the representative's fingertips. From the customer's point of view, it makes sense: Everything should be accessible from any computer. ServiceNow CSM connects company resources and insights, and automatically organizes and pulls together as many resources and relevant customer insights as possible. This is done in an intuitive way that maximizes the accessibility and utility of this data when it is retrieved by the customer service representative. By bringing in customer data into "one view," agents are able to offer the customer a more personalized experience, get a faster and more accurate resolution, and strengthen the long-term relationship with that customer.

3.5 ENABLE DIAGNOSIS OF ROOT CAUSES:

Traditional customer service involves solving customer problems one at a time, as if each case presented an entirely new and unique problem. The reality is there are many commonalities at the root of individual problems. Customers today expect that businesses will use other cases that come long before the customer's own to help diagnose and resolve their problem with speed and accuracy. In other words, it isn't just inefficient for a business to reinvent the wheel with each customer case; customers

also believe that institutional knowledge can and should be used to more efficiently resolve their issues. Taking time to find the root cause patterns and putting tools and processes in place for the agent to use when engaging with customers will speed resolution time, greatly improve customer satisfaction, and increase agent morale.

3.6 DELIVER CUSTOMERS INSIGHTS ACROSS THE FULL LIFE CYCLE:

Customers are increasingly cognizant that businesses collect a tremendous amount of data on them. The average company, however, is still struggling to meaningfully use this data to streamline and improve CX. A 2016 survey by Econsultancy found that 75 percent of companies believe understanding customer behavior over time is important to their growth, although just 12 percent reported having strong capability to generate such insights. You want to get ahead of your competition by thinking about how to mine customer data and all of your internal resources to generate insights that span the full customer life cycle.

Today's customers are intuitively aware of how advances in technology and CX strategy can drive the quality and speed of customer service delivery. They expect to see businesses make full use of these tools and approaches, and more importantly, to roll out CX solutions that are inspired by what customers value and prioritize. To exceed your customers' expectations consistently, create multiple contact center channels, focus on maximum speed to resolution, offer robust self-service options, utilize all potential resources across the enterprise, diagnose root causes, and deliver business insights across the full customer life cycle.

How confident are you in answering these questions?

- How prepared are you to articulate what your customers value and prioritize about CX?
- What does a fully integrated self-service knowledge base for your organization look like?
- On a conceptual level, what is your approach to diagnosing root causes of common customer issues?
- What is your strategy for ramping up capacity to generate customer insights using data analytics?

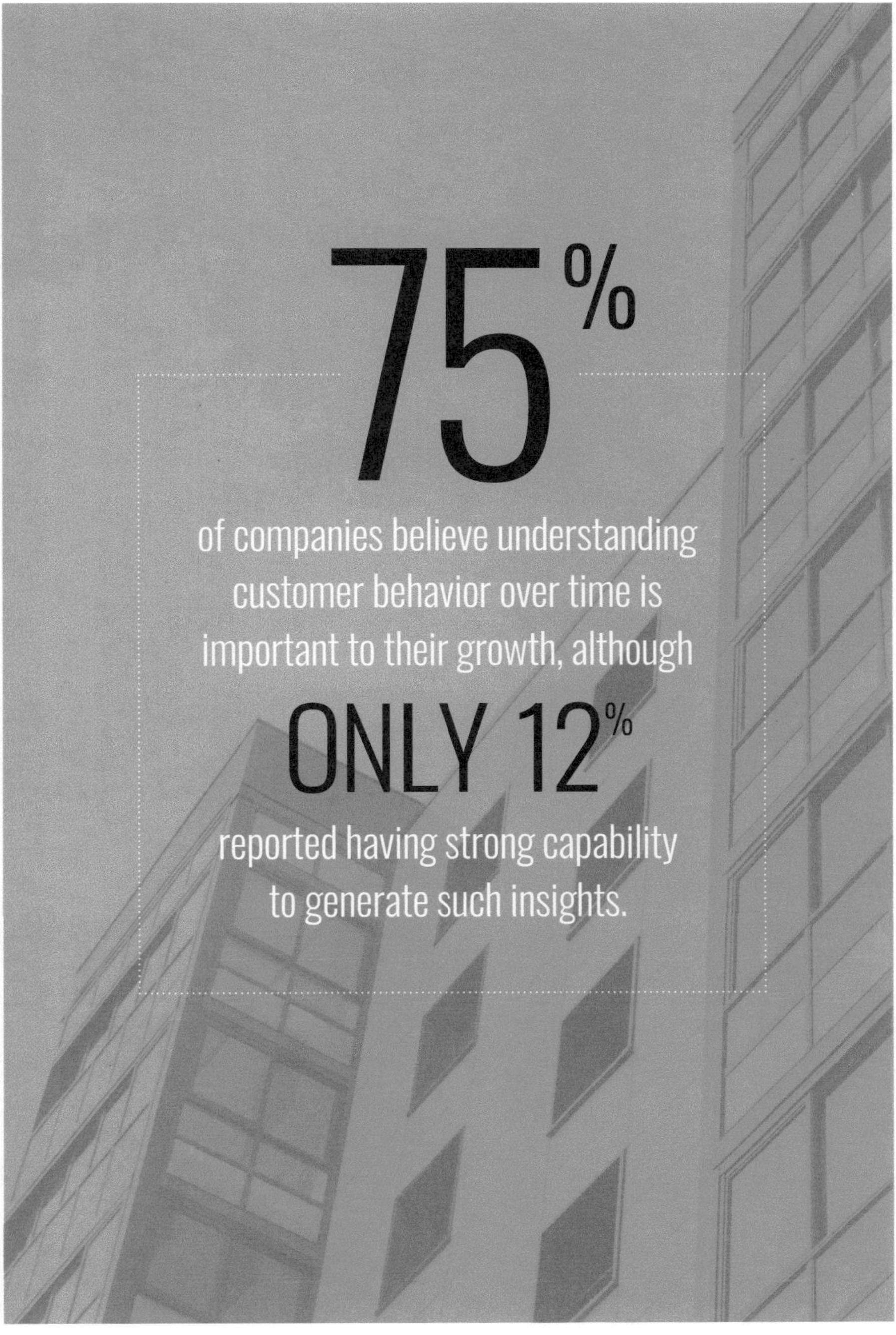

75%

of companies believe understanding customer behavior over time is important to their growth, although

ONLY 12%

reported having strong capability to generate such insights.

Pillar 4

Connecting Your Contact Center Team to Technology

MEET SHEILA

Sheila, an up-and-coming contact center director, inherited an operation wedded to outdated systems and workflows. When Sheila was promoted into her new position, she was shocked to discover that contact center representatives were still operating as they did 20 years ago: Representatives would take calls, manually type in relevant information about those calls, and store the records in a single, static database. More troubling, Sheila discovered that the contact center team had little desire to change. No one had ever put pressure on them to innovate, and they had no intrinsic desire to become more efficient. They were very much stuck in their comfort zone. How can Sheila effectively work with the team she inherited to drive the widespread change she envisions?

Sheila isn't alone in feeling apprehensive about introducing her contact center to a new CSM technology that is built on best-practices processes. The success of a rollout almost always comes down to the front-line team that is tasked with executing the technology vision. You want happy, technology-literate employees, because employees who are happy with their technology will radiate that happiness during their customer interactions and, in turn, make customers happy. As CSM modernization hits more and more contact centers across all industries, the need for a smooth, successful rollout has never been more important. By 2021, contact centers using a cloud-based CSM approach are expected to be a $16 billion industry, growing at an average compound annual rate of 24 percent, according to a 2016 study by MarketsandMarkets. To position customer service professionals to adopt and use CSM optimally, customer service leaders must orchestrate a rollout that meaningfully introduces—and strategically connects—front-line agents to the technology that powers CSM. Let's explore five essential tips for successfully connecting your customer service team to CSM technology.

4.1 ALIGN NEW TECHNOLOGY TO STRATEGIC BUSINESS GOALS:

More than most other people, contact center employees understand the downsides and drawbacks of technology. They interact everyday with customers who are frustrated and confused by technology; indeed, much of their time is spent helping customers fix complicated, technology-related issues. So it should come as no surprise that customer service teams are hesitant and skeptical about new technology, especially if they already are averse to change. That's why it's crucial that you help agents and management understand the value of CSM technology in the context of strategic business goals. CSM isn't simply the latest technology fad; it has strategic business value, and you want to explicitly share these goals with your team and track success along the way.

4.2 IMPLEMENT CSM FEATURES AND CAPABILITIES WITH A LONG SHELF LIFE:

CSM technology can be configured in countless ways to manage, store, and connect a multitude of different data points. As you assess how all of these features and capabilities might enhance your customer management workflows, you want to consider how your teams will be

> CSM is built with best practices in mind, which means that every time you make a tweak, it's critical that you know how that tweak impacts other areas of the platform.

affected as they use these workflows on a daily basis. You want to choose features with a long shelf life—features that will serve the organization well over the long term. Conversely, if you choose features that need to be constantly tweaked, refined, and replaced, you will be adding to agent frustration and increasing resolution time as agents struggle to stay on top of yet another change. Besides, your team's primary responsibility is to help customers, not to master complicated technology.

4.3 ADOPT YOUR CONTACT CENTER'S TERMINOLOGY AND WORKFLOWS:

As you configure your ServiceNow CSM application, keep in mind that CSM gives you great flexibility to customize it as you want. Before you dig in and begin to configure features and functionality sets, be sure you understand how one change ripples into another. CSM is built with best practices in mind, which means that every time you make a tweak, it's critical that you know how that tweak impacts other areas of the platform. You also don't want to implement your customizations in a vacuum. Take the time to outline your contact center's workflows and terminology before aligning them to your CSM platform configurations. This will simplify and streamline the process of getting your contact center team up to speed, which, in turn, will help the team identify and connect with the new CSM technology faster and easier.

4.4 START BY TRAINING INFLUENCERS:

When you're planning for CSM implementation across your contact center, start with a small group of beta testers and expand from there. The best folks for this job are the influencers within the contact center. These are the folks who motivate and inspire others, and whom others look up to; they can be found on both the front line and in leadership roles. Influencers play an instrumental role in ensuring all of their peers have a positive first experience with CSM and grow to embrace the technology. They also help to shape the future direction of your CSM platform as your contact center continues its journey to modernization. As you train your influencers on the new CSM platform, they'll provide feedback, and you can work through troubleshooting and system improvements with them. Ultimately, they'll motivate and inspire others as CSM is rolled out across the entire contact center.

4.5 CREATE A WORKPLACE CULTURE THAT REWARDS TECHNOLOGY ADOPTION:

To help contact center employees embrace CSM, you want to build a culture that rewards the sort of innovation CSM makes possible. This process starts with aligning job performance expectations to the strategic business goals of CSM implementation. You also can "gamify" the workplace with employee competitions that reward employees for adopting and achieving success with CSM. When you create these positive reinforcement measures, you are incentivizing employees to meaningfully connect with CSM technology.

The journey to a CSM victory involves people—lots of individual contact center employees who must adopt and embrace CSM for the technology to be effective. After all, happy contact center representatives translate to happy customers. Fortunately, you can take proactive steps to help your team feel connected and committed to CSM technology. This process starts with aligning new technology to strategic business goals, implementing CSM features with staying power, adopting terminology and workflows already used by contact center employees, training influencers first, and creating a workplace culture that rewards adoption.

How confident are you in answering these questions?

- How would you describe the readiness of your contact center teams to embrace new technology?
- What strategies for introducing technology to your contact center will resonate most with your employees?
- What measures can you put in place to ensure innovation gets prioritized and incentivized in your contact center?

Pillar 5
Utilizing a Customer "One View"

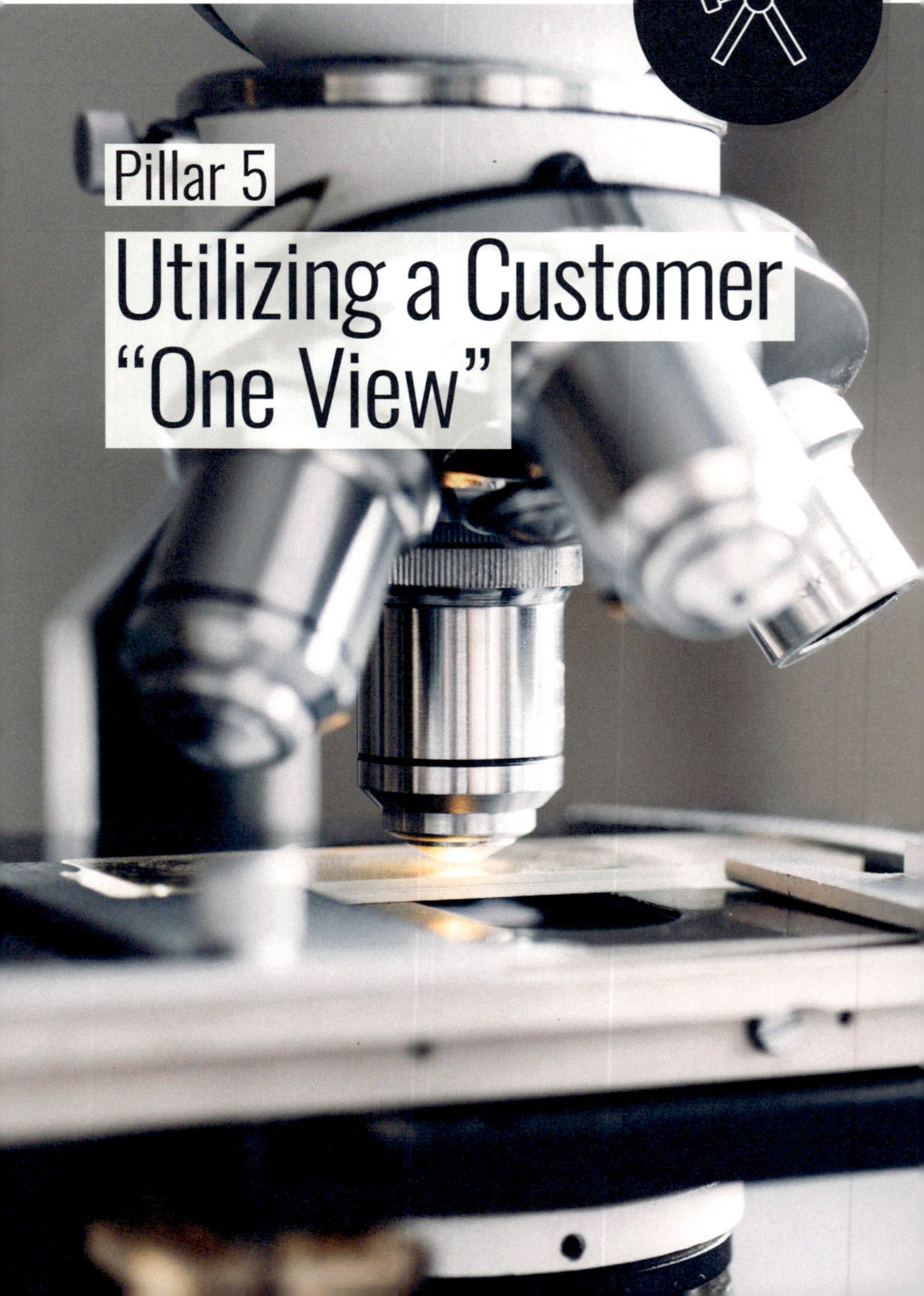

MEET ERIC

Eric, a senior vice president for customer experience, is like many of his peers in that he realizes the value of the "one view of the customer," even as he remains uncertain about how to design such an interface. Eric has read extensively about how contact center representatives benefit from being able to see key customer data displayed on a single interface—specifically, how this information makes each customer interaction personalized and customized, and ultimately creates more brand-loyal customers who spend more and seek out long-term relationships with a company. Eric understands how a single customer view works in theory, but doesn't know how his contact center will get there operationally. As is the case with many contact centers, the information Eric needs is stored in static, decades-old databases, and he knows there are no links among CRM data, customer purchase records, social media engagement records, and marketing campaign history. In fact, there isn't even a department or team tasked with running analytics on this data. And even if such insights were available, Eric's contact center representatives aren't trained in how to use data and analytics to drive CX. How can Eric find success in implementing a single customer view for his contact center?

Eric's struggles aren't unique. When asked about the value of a "one view of the customer," virtually every company agrees that a single customer view is important for the success of the business. But only 24 percent of businesses report actually having a single customer view, according to a recent survey by Experian Data Quality. The reality is that many organizations struggle with inconsistent data collection, data quality challenges, and an inability to meaningfully link customer data. Fortunately, none of these hurdles are insurmountable. Let's review the four most important tips to remember when developing your "one view of the customer" and optimizing CX.

5.1 LINK INFORMATION STRATEGICALLY TO DELIVER INSIGHTS:

The biggest challenge when building one view of the customer is to link disconnected customer data meaningfully and strategically. You want to take the time to discover all potential relationships within your existing databases, so you can think critically about which linkages would most improve how contact center representatives relate to and engage with customers. You want to cast a wide net: Consider geographic, demographic, and industry-specific data. Think about unconnected data collected from across multiple channels, from social media to direct marketing to customer satisfaction surveys. Have open-ended conversations with your front-line contact center team to understand what information and relationships they would find most useful during their day-to-day customer interactions.

5.2 INTEGRATE ENRICHED, CONTEXTUALIZING DATA:

The data your own organization collects is crucial for building one view of the customer, but it's not the only data you need. By enriching your data with third-party market intelligence, you can draw even more relevant insights that help you deliver more seamless, engaging customer service. For example, enriched data sets can provide important contextualizing information about the general life cycles and warranties of products and services your organization sells. Enriched data help contact center representatives to answer customer questions more accurately and comprehensively, as well as to offer up more engaging insights.

5.3 ENSURE DATA IS CLEANSED AND STANDARDIZED:

For contact center representatives to draw on the data in a single customer view to jump-start conversations, representatives need to be able to count on the data's accuracy. That's why it's crucial for all data analytics capabilities to include quality-control measures that help weed out erroneous data and that prevent it from entering the system in the first place. Ideally, automated quality-control checkers should be set up to reject data that can't be read and understood by data analytics tools later.

5.4 USE PREDICTIVE ANALYTICS TO OPTIMIZE CX:

In addition to providing data about customers, you also can predict what future customer data will look like and use it to engage in more precise forecasting and long-term planning. Predictive analytics capabilities allow organizations to use real data to detect patterns and estimate future trends. For example, you could use predictive analytics to project customer selling habits, how customers will respond to certain types of marketing campaigns, and the rising and falling of inventory and resource needs over the course of the year.

24%

of all businesses report that they've experienced an IT project failure within the past year.

A "one view of the customer" is optimized when it integrates key data sets to measurably move the needle on long-term customer engagement, sales, and profitability. With a single customer view, contact center representatives can consistently do their jobs more effectively and efficiently, saving time and money. To build this integrated system, you want to link information strategically to deliver insights, to integrate enriched data to give your data more relevance and meaning, to ensure data are standardized and cleansed, and to use predictive analytics to optimize CX.

How confident are you in answering these questions?

- What should a 360-degree view of your customers look like?
- What is your vision for more effectively integrating your customer data to increase sales?
- How can you bring standardization to your data and ensure it gets systematically cleansed?

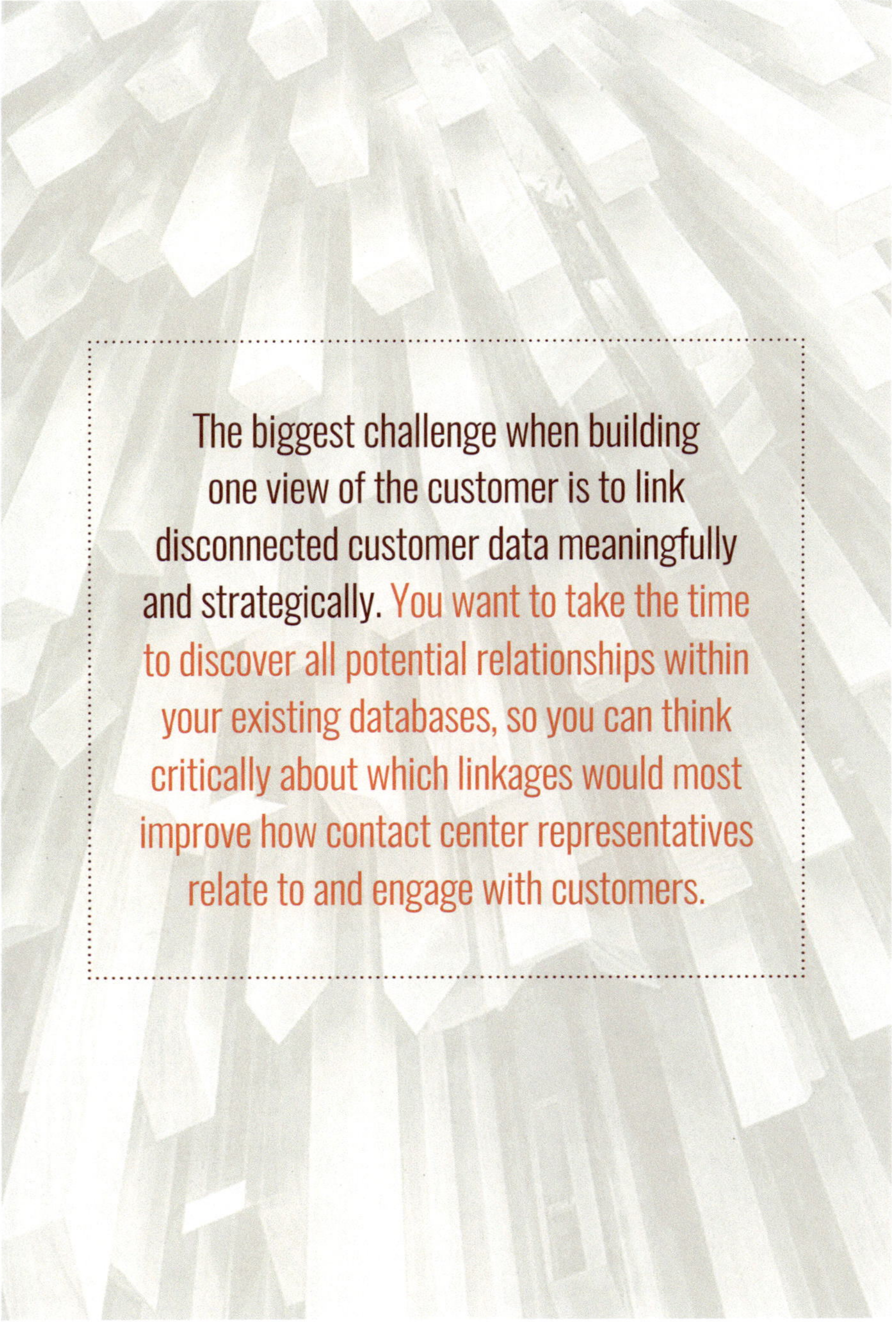

> The biggest challenge when building one view of the customer is to link disconnected customer data meaningfully and strategically. You want to take the time to discover all potential relationships within your existing databases, so you can think critically about which linkages would most improve how contact center representatives relate to and engage with customers.

Pillar 6

Instilling Confidence and Pride in Customer Service Professionals

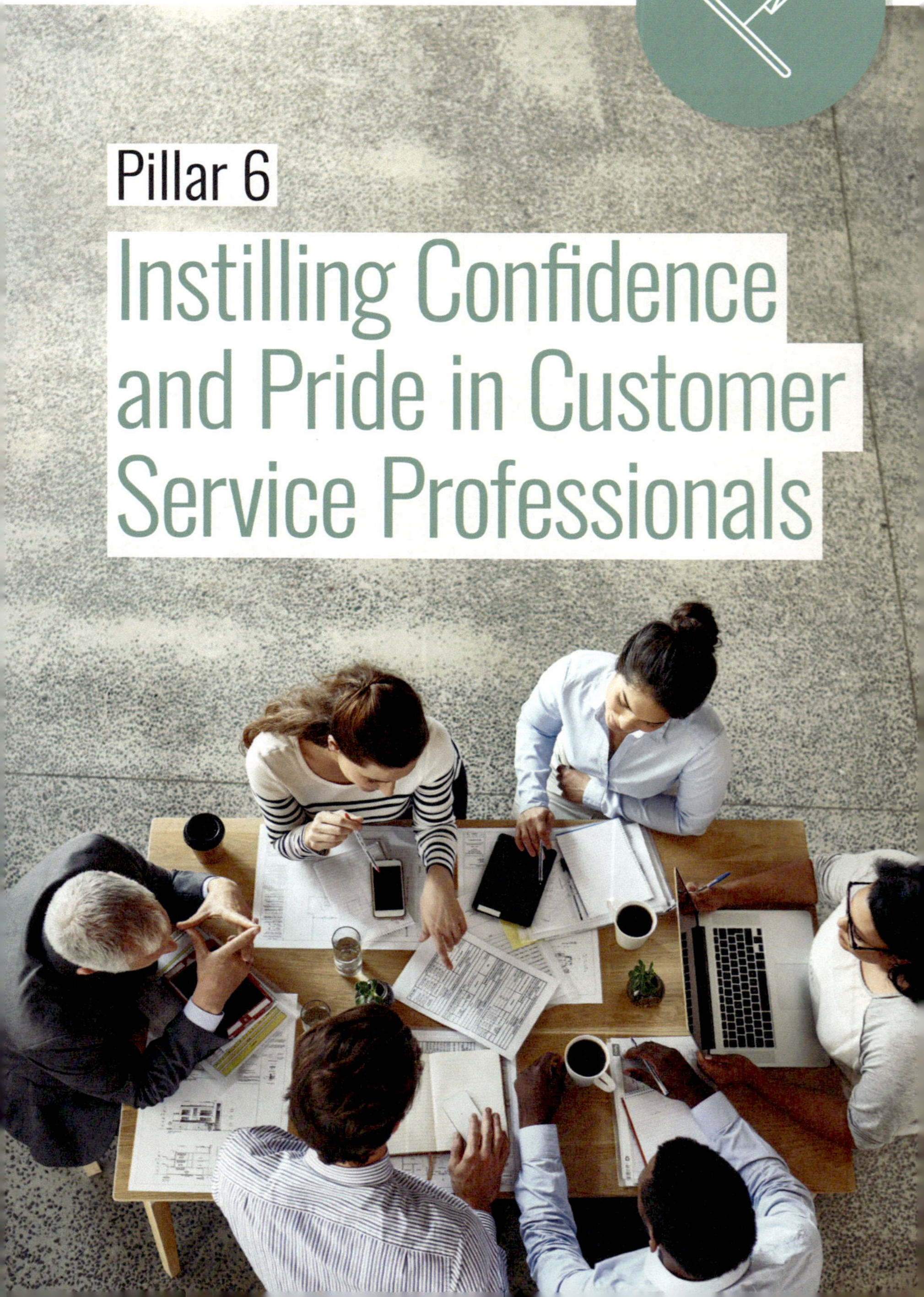

MEET EMMETT

Emmett is a veteran customer experience leader with a big problem on his hands: His contact center agents are disengaged from their jobs and are frustrated about their inefficient, outdated computer systems. Emmett is planning to modernize the entire contact center operation with ServiceNow CSM, but he's afraid that with such low employee morale and dismal engagement levels, his most important and transformative project could be derailed. How does Emmett create a culture where his contact center is confident and proud of its customer service abilities?

ONLY 32% of U.S. employees are enthusiastic about and committed in their jobs and workplaces.

What Emmett is picking up on is a very real phenomenon: Employee engagement matters, both from a customer-focused perspective and from a strategic business perspective. To be successful, contact center representatives need to feel confidence and pride in their jobs and toward their employer. Unfortunately, disengaged employees are pervasive across all industries. Only 32 percent of U.S. employees are enthusiastic about and committed in their jobs and workplaces, according to a 2016 Gallup poll. If your contact center employees are representative of these dismal statistics, you've got your work cut out for you. Fortunately, there are proactive steps you can take to re-engage your contact center employees as you prepare for CSM implementation and optimize how you position a new platform to them. Let's explore five essential strategies for instilling confidence and pride among the team ahead of your contact center transformation.

6.1 TEACH EMPATHY TO YOUR EMPLOYEES:

A common reason that contact center representatives lose engagement in their jobs is because they lose sight of what it's like to be in their customers' shoes. The best way to combat this problem is to teach empathy. Remind employees of how anxious and frustrated customers can feel when they are having a problem with something that is important to them. Help employees understand that everyone wants to be treated special, to receive personalized service, to be surprised and delighted by a customer service experience.

6.2 IMMERSE YOUR EMPLOYEES IN COMPANY VALUES AND STRATEGY:

The grind of resolving customer issues day in and day out can take its toll on your contact center team. They may entirely lose sight of why their job is so important, and they can quickly forget the critical role they play in moving the needle on customer engagement, satisfaction, and sales. The solution is to constantly remind your contact center team of how their work relates to overall business values and strategy. With every memo that your contact center receives, with every training and presentation they attend, you want to immerse your employees in corporate strategy and connect them in meaningful, personalized ways to company values. Be sure, too, to showcase examples of employee success stories that have supported corporate strategy. This will show that you appreciate their dedication to aligning their mindset and actions for the betterment of the company, your customer, and their career.

6.3 EMPOWER YOUR EMPLOYEES TO MAKE DECISIONS:

Contact center employees struggle every day to find appropriate, personalized solutions for customers. As they make split-second decisions about how to resolve customer issues, you can support them by cutting out

Showcase examples of employee success stories that have supported corporate strategy. This will show that you appreciate their dedication to aligning their mindset and actions for the betterment of the company, your customer, and their career.

the red tape and empowering them to make as many decisions as possible. For example, you want to allow them to refund and credit customers without supervisor approval, and to use their institutional knowledge and skills to make recommendations and offer advice of all kinds. The less restrictions and rules that you place on contact center employees, the more they feel empowered and engaged in their jobs. You also want to develop ongoing training based on common customer requests and issues to help agents feel confident in split-second decision-making. When employees are confident in their decision abilities, they will not only feel empowered, but also more satisfied with their job.

6.4 CREATE RECOGNITION PROGRAMS FOR CX STANDOUTS:

Employees respond to recognition for doing exceptional work. You can incentivize this behavior by creating recognition programs that recognize not only representatives with high customer satisfaction scores, but also representatives who are early adopters of the CSM application and who achieve quantifiable successes as a result of using it. Recognition doesn't just involve simple certificates and rewards; you also can create games and quizzes that recognize the top finishers, and use leaderboards to foster some friendly competition and bragging rights.

6.5 ASK YOUR CUSTOMER SERVICE EMPLOYEES FOR FEEDBACK:

Your contact center employees work on the front lines every day and know what processes and policies work. Especially as you begin developing your vision and strategy for CSM implementation, you want to involve your front-line CX employees in the process. Think of your contact center employees as the most insightful stakeholder group you could imagine. When they see that you take their feedback seriously and make adjustments based on what they recommend, they will become more engaged and committed to their work.

Contact center teams are an organization's first and primary point of contact for establishing a positive brand experience, both among internal department teams and among outside customer communities. You want to do everything you can to support and engage your customer service team so that they become effective brand ambassadors who come across as confident, knowledgeable, and helpful. The journey to instilling confidence and pride in your agents begins with teaching empathy, immersing employees in corporate values and strategy, empowering employees to make decisions and provide feedback, creating recognition programs for CX standouts, and asking your user base of agents and management for their feedback, particularly during the transition to CSM.

How confident are you in answering these questions?

- What strategies could you deploy to instill confidence and pride in your contact center team?
- Where are the gaps between how your contact center team relates to customers and how this team should relate to them?
- How can you systematically draw out meaningful feedback from your contact center team?

Pillar 7

Creating Consistency in Omnichannel Delivery

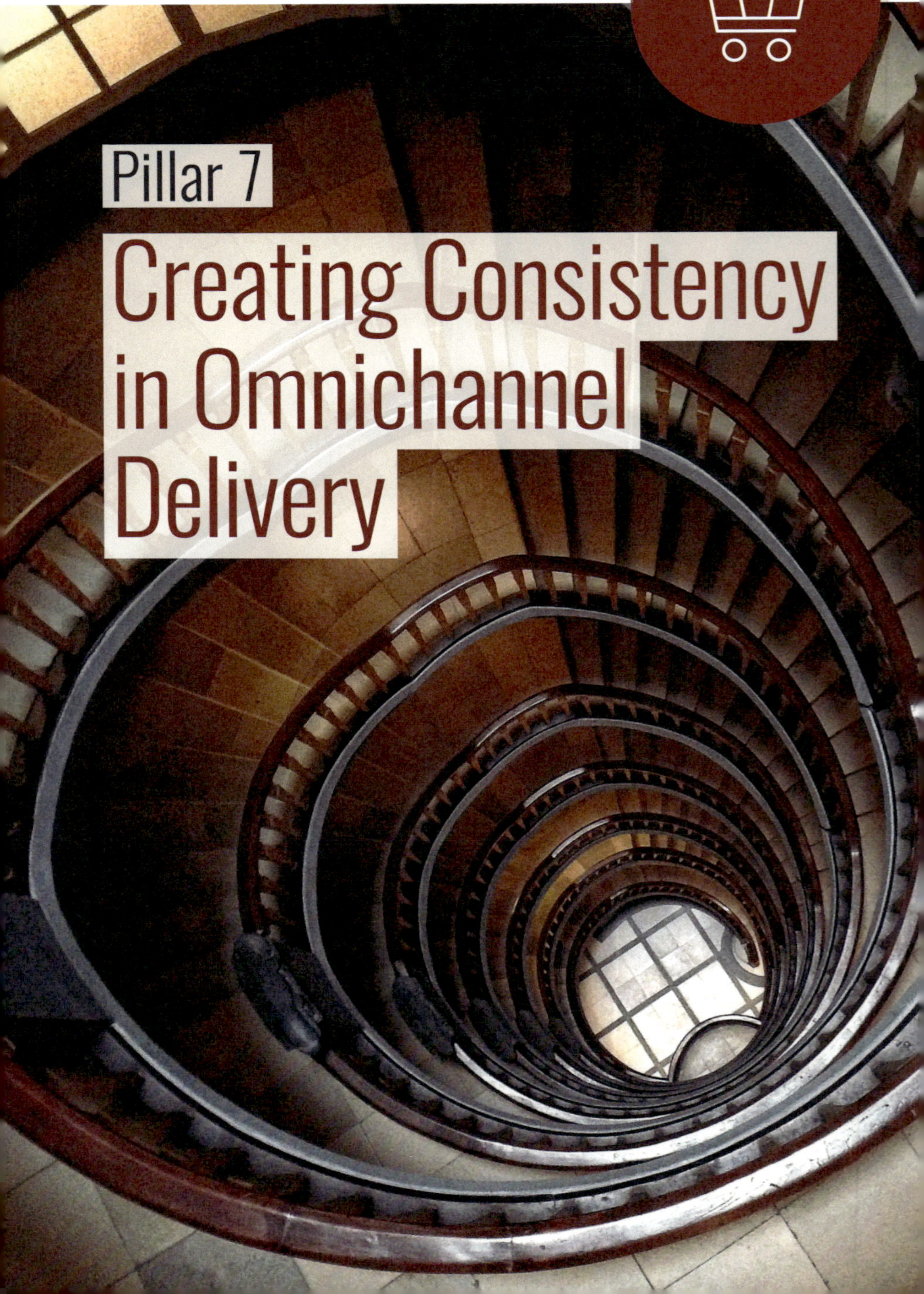

MEET STACY

As head of customer service initiatives, Stacy understands the critical importance of a multi-channel approach to CSM. She recognizes that customers expect to have full flexibility when communicating with her organization—whether they choose phone, email, web chat, text message, social media, or knowledge bases. However, as is the case with most organizations, the multi-channel delivery of customer service at Stacy's company has been developed in a very ad-hoc, non-centralized manner: Stacy oversees a social media team that operates largely in a silo, the CRM system that her contact center is using isn't linked to the customer support system, and the web team has developed its own separate processes and protocols for responding to and engaging with customers. Stacy wants to take on a massive overhaul of customer service delivery across all channels, and she wants to do it at the same time as she oversees implementation of ServiceNow CSM. How does Stacy set the wheels in motion to unify all customer-facing channels in an orderly fashion? And how does she avoid unnecessary risk to CX or agent morale as she works toward these goals?

Contact center leaders like Stacy are correct in prioritizing omnichannel delivery of customer service. A 2016 survey by Statista found that just 58 percent of all customer interactions in contact centers are handled over the phone nowadays. That means about two out of every five customers are forming an impression of companies using communications channels other than traditional telephone interaction. Like all customers, these customers expect consistency and expediency in their service experience, and a personalized, quality solution that fully resolves their issue. To ensure contact centers can reliably deliver high-quality CX regardless of channel, it is important for CSM leaders to take proactive steps to unify disconnected systems and processes. Let's explore four essential strategies for unifying contact center channels to create consistency and improved service delivery.

7.1 LOG EVERY CUSTOMER INTERACTION IN A SINGLE DATABASE:

Contact centers are generally good about logging records of phone calls and perhaps even emails and web chats, but what about social media interactions? Furthermore, even if you have the ability to log every customer interaction across all contact center channels, all of these records ultimately must be centralized in one database to be useful. The reason is simple: When customers make subsequent contact, they automatically assume that the company can instantly pull up a complete history of all interactions. Hence, if your system doesn't do this, you've already disappointed a customer you are working to build a relationship with and effectively support over the full customer life cycle. What makes CSM so powerful is that it brings all customer interactions together, and makes them easily accessible from a "one view of the customer" perspective, providing critical context and insights about the customer. And with this centralization of records also comes other capabilities. For example, if your contact center representative asks for a serial number and the customer doesn't know how to access it, you want to be able to send your customer an article on how to retrieve this information in real time.

7.2 CREATE UNIVERSAL EVALUATION METRICS FOR ALL CHANNELS:

Because many contact center channels have evolved independently of one another, contact center leaders often have developed ad-hoc, non-centralized metrics to gauge customer satisfaction and evaluate the job performance of agents. This inconsistency in evaluation metrics for contact center channels has led to uneven delivery of customer service across channels. Thus, it's important that you ensure that evaluation metrics are the same across all channels. As these metrics are refined over time, the entire contact center—regardless of channel—will be able to move in lockstep toward improving CX.

7.3 USE PRESENCE INDICATORS TO EFFICIENTLY ACCESS EXPERT HELP:

One of the most important metrics of any customer interaction is to provide first-contact resolution. However, as contact centers become more specialized and fragmented by channel, an unfortunate byproduct is that agents become less likely to be able to resolve every issue on their own. Presence indicators within the contact center can help agents in these cases find a more rapid solution—without the customer even being fully aware that the representative needs to escalate the customer's inquiry. With presence indicators, any representative can pinpoint exactly which subject-matter experts are available at any given time to help a contact center representative, and work in real time with that expert to identify an optimal solution. In this manner, contact centers can consistently achieve their first-contact resolution goals with speed, accuracy, and confidence.

7.4 INCENTIVIZE CROSS-CHANNEL PROFICIENCY AMONG EMPLOYEES:

No matter how much effort you put into standardizing delivery of customer service across channels, your success in creating omnichannel consistency ultimately boils down to the people who implement it on a daily basis. In many contact centers, agents tend to be pigeonholed into working within one channel only, and may become comfortable and complacent. The way to broaden your employees' perspective—and create greater consistency in the process—is to incentivize cross-channel proficiency. Representatives who know how to deliver customer service across multiple channels are going to be more knowledgeable, more motivated, and more committed to their jobs. And because they're working across channels, they'll naturally be able to close the gaps you're facing in bringing consistency to omnichannel service delivery.

As you standardize delivery of customer service across all channels, you'll find that not all channels are as resource-intensive as others. In particular, your customers are likely to feel completely satisfied communicating with you through social media or turning to a self-service portal for help—if the quality is there. Over time, you want to track the portions of customers who use your various contact channels. The cost per interaction varies dramatically by channel, which means that as you learn to effectively serve your customers through less expensive channels, your operational costs will go down and you'll be able to maintain your competitive edge. To create high-quality consistency across your channels, you want to start by logging every customer interaction into a single database, creating universal evaluation metrics for all channels, using presence indicators to efficiently access expert help, and incentivizing cross-channel proficiency among your contact center agents.

How confident are you in answering these questions?

- How much CX inconsistency is there across channels in your contact center?
- What strategies could you use to gently push more customers to less expensive contact channels?
- What evaluation metrics can you apply to evaluate CX delivery across all of your contact channels?
- How can you effectively incentivize your contact center team to become proficient across multiple channels?

Pillar 8
Adopting a Mobile-First Mindset

MEET MARC

Marc is a senior vice president of customer integration, responsible for delivering seamless customer service experiences. Like many organizations in the industry, Marc's company wants to become better at targeting and appealing to millennials. They represent the biggest growth opportunity for his company, even as millennials presently account for only a small fraction of the company's total sales. Marc's dilemma is that his contact center was set up decades ago and has evolved only incrementally over the years. Indeed, when the contact center leaders who report to Marc boast that they've kept up with "mobile," what they're actually referring to is the basic technology infrastructure upgrades they've made to the contact center to support integration with mobile devices. They haven't, however, adopted a truly mobile-first mindset that focuses on optimizing the customer experience for mobile devices. As Marc prepares to modernize his contact center operations and implement the ServiceNow CSM platform, he wants to use a mobile-first mindset to guide the implementation. How should Marc apply the mobile-first mindset to reimagine his contact center operations?

Marc's challenge is all too common among companies whose contact centers came of age in a different era. Today's customers, especially younger generations, rely increasingly on their mobile devices for information, entertainment, and record-keeping—both in their personal lives and in business. A whopping 89 percent of Americans check their smartphones throughout the day, and 57 percent check it at least once an hour, according to a recent study commissioned by Bank of America. Like all customers, mobile-first customers have high expectations when they make contact with businesses: They assume they'll find high-quality interfaces, get the information they're looking for almost instantly, and be met with knowledgeable, helpful customer service professionals who can effortlessly find them a customized solution. For a business to succeed in this mobile-first world, the contact center and the CSM strategy itself must be aligned to a mobile-first mindset. Let's explore four essential strategies for optimally applying a mobile-first mindset to reimagine your operations and put your contact center on a path to sustainable success.

8.1 FOCUS ON CONTINUOUS IMPROVEMENT ACROSS ALL CHANNELS:

Mobile devices, which include both smartphones and tablets, have ushered in a new era of communication focused on streamlined, intuitive interfaces and responsive design. Gartner estimates that market demand for mobile app development services is growing as much as five times faster than businesses' IT capacity to deliver them. To ensure that customers have a positive experience as you roll out and enhance mobile-oriented communication channels, you want to set an expectation of continuous improvement. Teams should be assigned to routinely test and fine-tune the CX capabilities of mobile devices, and CX-related feedback from customers should be actively solicited and methodically reported to keep these channels functioning optimally.

8.2 ENSURE CONTACT CHANNELS ARE ACCESSIBLE AND PROMINENT:

You can design the most technologically advanced contact center channels in the world, but if the average customer has trouble finding them and using them, you won't achieve the outcomes you seek. Just as we all understand the importance of displaying a toll-free number prominently, we must make mobile-friendly contact channels equally as prominent and accessible. When your customers go searching for their preferred channel to contact you, they should be able to find it instantly, and they should intuitively be able to figure out how to use it and get exactly what

Gartner estimates that market demand for mobile app development services is growing as much as

FIVE TIMES

faster than businesses' IT capacity to deliver them.

they need without delay. For example, when you're writing and publishing knowledge base articles, you should have quality-control measures in place to ensure the articles are SEO-optimized, properly organized and indexed, and interlinked with all other relevant self-help resources.

8.3 USE CONTACT CHANNEL ANALYTICS TO PERSONALIZE CX:

With the mobile revolution has come unprecedented opportunities to mine big data for insights into customer preferences and spending habits, values and priorities, and common pain points. CSM enables customer data to be aggregated from across all contact center channels, so that contact center leaders can make better-informed decisions about how and when to communicate with customers (and, just as importantly, how and when not to communicate). For example, a contact center agent comes across as extremely knowledgeable and competent when he or she can say, "I see you ordered Item X on April 15, and you called three days ago to inquire about your order status. We just received a bulletin yesterday from our supplier letting us know about a four-day delay, and we can make up some of this lost time by automatically adding expedited shipping to your order—all at no additional charge to you, of course."

8.4 PRIORITIZE SECURITY OF CUSTOMER DATA:

As you take advantage of mobile capabilities to collect big data on your customers, it's crucial that you implement new security measures to protect that data. Customers have an expectation—no matter how unrealistic—that every business they interact with can maintain their privacy and keep their data secure 100 percent of the time. You don't want to lose your customers' trust, and you definitely can't afford to lose the sales. It's far more expedient to proactively invest in managing customer data security and take steps to safeguard their privacy. The last thing your brand needs is to be the headline story on CNN because of an embarrassing data breach!

With the mobile revolution has come an unprecedented ability to leverage customer insights and analytics to drive new sales opportunities. At the same time, adopting a mobile-first mindset requires an upfront investment—infrastructure upgrades, new technology adoption, and a shift in strategy. To adopt a mobile-first mindset and put your contact center on a path to success, you want to focus on continuous improvement, make your contact channels accessible and prominent, personalize the customer experience using analytics, and prioritize the security of your customers' data.

How confident are you in answering these questions?

- How prepared is your organization to embrace a mobile-first mindset?
- What steps could you take to immediately raise the access to your contact center channels?
- How could you begin mining the customer data you already collect to deliver key customer insights?
- What additional customer data could you collect to deliver even more business insights?
- What processes and systems are in place to protect the security of your customer data?

Pillar 9

Utilizing Big Data Analytics to Enhance CX

MEET CHARLES

Charles, the longtime head of customer experience for his company, is in the middle of CSM implementation—and he's fearful he won't be able to demonstrate a strong return on investment. He has prepared a strategic vision and an implementation roadmap, and he's gotten widespread buy-in and support from C-suite leaders and front-line contact center representatives. Still, Charles knows the C-suite has big expectations for CSM, due in no small part to how much Charles himself has talked up CSM. As Charles embarks on his implementation journey, he feels nervous about how he'll guarantee a strong return on investment. He's been reading up on big-data analytics, which involves focused, computer-driven analyses of customer data to drive greater customer spending and loyalty. The more Charles learns about data analytics, the more opportunity he sees—and the more trepidation he feels. How can Charles apply big-data analytics principles and best practices to optimally transform customer service operations?

Charles' struggle to use big-data analytics to better understand his customers is a universal challenge. Companies all over the world are struggling to personalize customer interactions meaningfully, in ways that will reliably and measurably move the customer satisfaction needle. A 2014 Forrester research study found that the single biggest struggle faced by marketing is understanding customer interactions across contact center channels; 46 percent of marketers reported it as a major challenge, placing it ahead of other significant obstacles like securing adequate staffing and a sufficient budget. While big-data analytics can be complex, there are certain unifying principles around which to build a robust analytics program that can enhance CX and drive profitability. Let's explore five principles for how to effectively use big-data analytics to enhance CX.

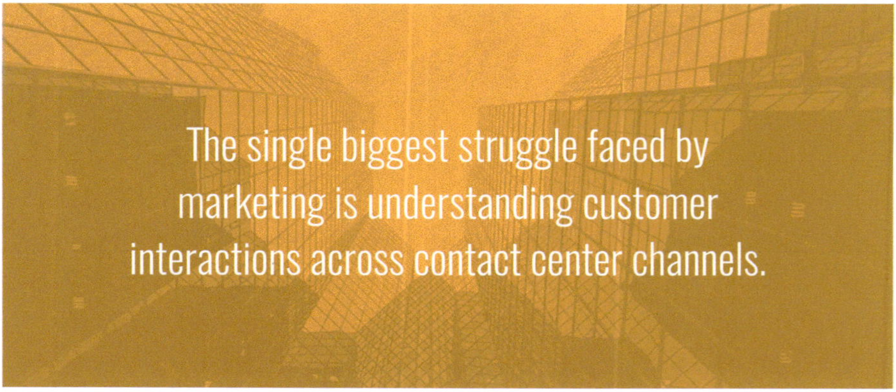

9.1 IDENTIFY FRUSTRATED, DISENGAGED, AND HYPER-ENGAGED CUSTOMERS:

Customer data can be parsed and analyzed in all sorts of ways to generate endless big-data insights into who your customers are and what they want from your company. But the most important pieces of information to glean from big-data analytics are how to pick out three types of customers: your frustrated customers, your disengaged customers, and your hyper-engaged customers. Frustrated customers are those at risk of stopping doing business with you, and for which time is of the essence to prevent them from becoming part of your churn. Disengaged customers are those who aren't using the products and services of yours that they're paying for—or at least not to their full potential; these are the customers you need to bring back into the fold to keep them happy and enthusiastic about maintaining a relationship with you. Lastly, hyper-engaged customers are those who are making full use of your products or services and who are so satisfied with your company that they would be glad to hear from you; these are the customers who are predisposed to consider enhanced service levels and purchase additional products. One important note here: These classifications apply just as readily to your employees, who are essentially your internal customers.

9.2 TIE ANALYTICS TO SPECIFIC ACTIONS THAT CAN BE EXECUTED CONSISTENTLY:

Big-data analytics enable you to produce not just customer insights, but also to trigger specific actions based on what that data tells you. For example, if you sell SaaS software, you could use analytics to automatically alert you to all customers who haven't logged in over the past month. If you ship products to your customers, you can install triggers and flags that tell you that, for example, 10 percent of a particular customer's packages are being packed wrong. Conversely, if there is login activity on every single license that your customer has purchased, you could use analytics to automatically notify your sales team that this customer is ready for an upgrade. Big-data analytics enable such actions to be executed consistently, ensuring you take a steady approach to sales and customer service that does not let any customers slip through the cracks.

9.3 ENSURE KEY PERSONNEL GET ASSIGNED TO HIGH-VALUE, AT-RISK ACCOUNTS:

With ServiceNow CSM, customers can be automatically routed to the individuals and teams that are most likely to get them to a satisfactory resolution. There is no customer more important than the high-value one at risk of jumping ship. In the case of CSM, you want to use its real-time data analytics capabilities to ensure your high-value customers get connected to your best contact center agents or other support tiers—the ones who specialize in retention.

9.4 FOCUS ON IMPROVING THE SPEED AND QUALITY OF CUSTOMER RESOLUTIONS:

As you configure CSM to take advantage of your big-data analytics, you want to maintain focus on gleaning customer insights that will improve the speed and quality of customer resolutions. That means being able to serve up important information in a context that helps your contact center agents personalize each interaction and rapidly diagnose and resolve the customer's issues. Big-data analytics allow you to standardize this information, so you can count on it to add value to every customer interaction, especially as you build rapport with long-term customers over time.

9.5 USE ANALYTICS TO PREDICT CUSTOMER BEHAVIORS AND CHANGES:

Analyzing data after the fact is incredibly valuable, but big-data analytics isn't limited to the insights you draw from existing data. You also can use your data to make informed predictions of future customer behaviors and business trends—a field known as predictive analytics. Using sophisticated algorithms, you can estimate, for example, the increase in your sales if you double your advertising budget by 50 percent, or the price point at which you will start to see a drop-off in new customer leads. Think of these as "what if" predictions. Perhaps the best part about predictive analytics is a phenomenon called machine learning, in which your algorithms get better at making predictions as you feed them more data. What this ultimately means is that you're able to focus your limited resources on the customers who need the most attention and represent the highest value to your company.

High customer satisfaction scores are elusive for companies that struggle with consistency in the delivery of customer service. The reality is that stellar customer service requires extreme consistency and insightful customer data—qualities that can only be fully realized only with the help of big-data analytics. To optimize the use of analytics to drive memorable CX in your organization, you want to prioritize identifying frustrated, disengaged, and hyper-engaged customers. You also want to tie analytics to actions that can be executed consistently, ensure high-value customers get assigned to your most talented contact center agents, improve the quality and speed of customer resolutions, and use analytics to predict customer behaviors and changes.

How confident are you in answering these questions?

- How prepared are you to pinpoint who your frustrated, disengaged, and hyper-engaged customers are?
- What standard actions are you prepared to implement in response to what big-data analytics might reveal to you?
- What sorts of customer behaviors would you like to be able to predict to optimize CX?

Pillar 10

Innovating to Maintain and Strengthen Customer Loyalty

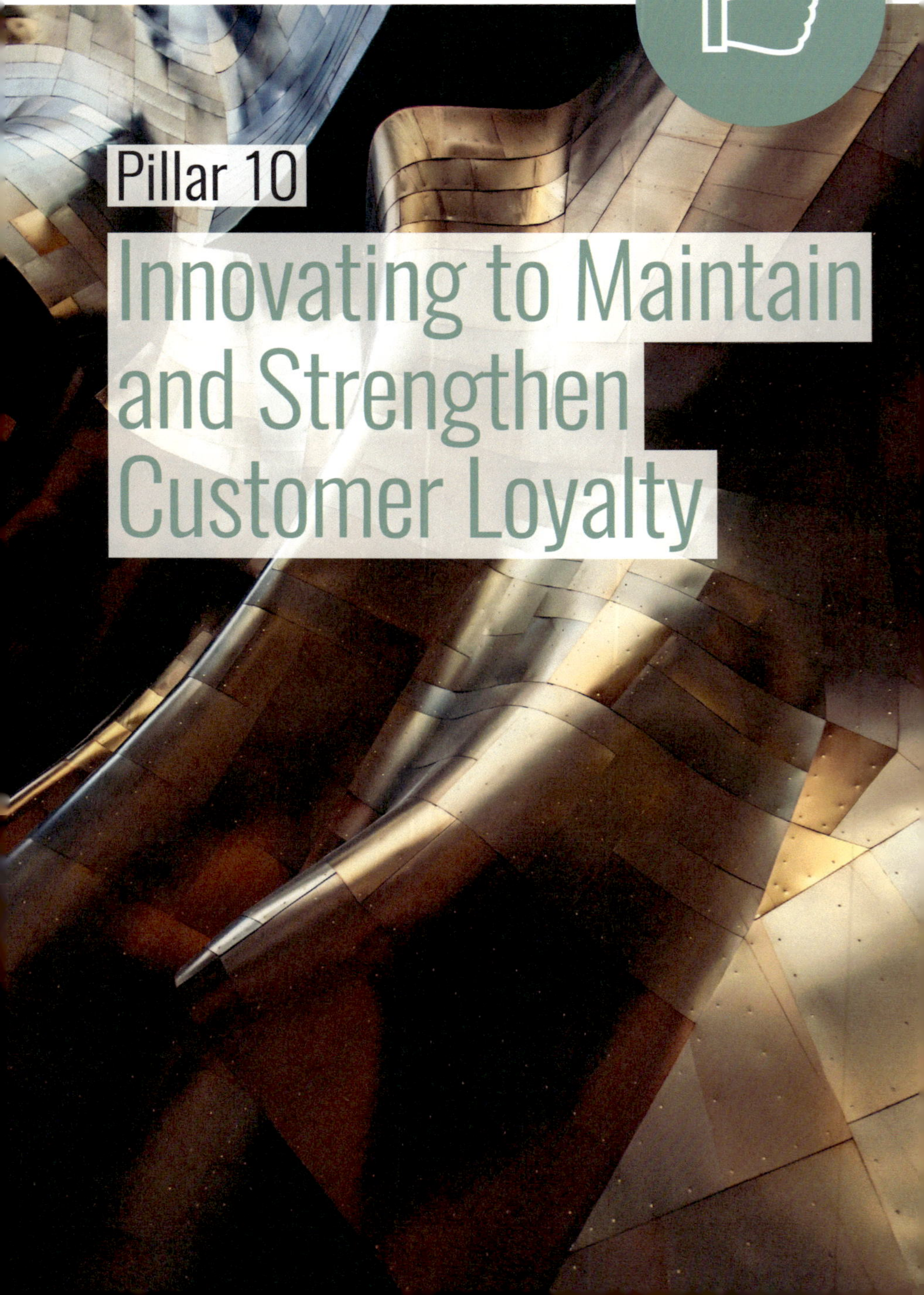

MEET MARY

Mary is a newly installed senior vice president for customer experience. Mary's predecessor, who was forced out, was a strong day-to-day manager who ran a tight ship, but he wasn't a visionary and struggled to execute a planned modernization of the contact center. Although CSM is now in place across the contact center to keep its core customer service capabilities running, Mary has already seen that there's no long-term vision to maintain a competitive position in the marketplace and drive new profits. She also observes that CSM hasn't been optimally implemented, and there's no internal momentum to innovate and work toward continuous process improvement. Mary intends to turn around this ship, and she wants to do it in a focused, strategic way that will infuse the contact center with a much-needed spirit of innovation. How should Mary focus her efforts to foster a culture of continuous innovation that maintains and strengthens customer loyalty over the long term?

The importance of continuous innovation cannot be overstated. Customers from all walks of life care about innovation in the companies they do business with, and not just the type of innovation that results in better products and services. A 2015 study from the market research firm Lab42 found that 84 percent of consumers say it is somewhat or very important that the company they buy from is innovative, a finding that holds up even among "dull" companies like insurance and gas. Lab42 noted that customer service innovation was specifically called out as a factor that shapes how customers perceive innovation in business. Fortunately, there are certain strategies you can implement that will play an outsized role in driving innovation within contact centers in the coming years. Let's explore four essential strategies for innovating in the CX space to maintain and strengthen your long-term customer relationships.

10.1 IMPROVE YOUR ABILITY TO INITIATE CONTACT WHEN YOU IDENTIFY PROBLEMS:

Contact centers have traditionally operated in a reactive mode, where they learn of problems when customers make contact to ask for help resolving the problem. By the time customers make contact, they've already unexpectedly been inconvenienced. CSM technology is enabling contact centers to increasingly operate in a proactive mode, where they can make the initial contact with customers to let them know of common problems in real time. For example, if you know of an outage that will affect 20 percent of your customers, you are far better off proactively notifying all 20 percent of them in advance about what's coming—instead of waiting on them to individually contact you as the problem begins to manifest. You'll not only reduce the workload of your contact center team, but you'll also have less frustrated customers.

10.2 TAKE ADVANTAGE OF THE INTERNET OF THINGS TO GAIN ENHANCED INSIGHTS:

The Internet of Things (IoT) has enabled more devices to be connected more often to more companies, which, in turn, are collecting more data about how customers are using and engaging with products and services. Amazon is a global pioneer in using IoT to enhance its already first-rate customer service. With the introduction of the Alexa voice service, Amazon is on the cusp of yet another CX revolution, where customers willingly share their needs, preferences, and behaviors with Amazon to—ironically enough—enhance their personal CX with Amazon. You want to be continually thinking outside the box to identify opportunities to take advantage of IoT devices—from smartphones to tablets to smart speakers to cloud-connected enterprise machines. Not only do IoT-connected devices give you a treasure trove of new insights into your customers' usage habits, but you can build remote diagnostic capabilities into your products, enabling you to be alerted when something goes awry with one of your products or when a customer isn't using your product as intended. With such capabilities, you might be able to fix problems without ever dispatching a site technician, or proactively reach out to a customer to help them resolve issues and get the most out of their product.

10.3 BOOST YOUR SELF-SERVICE OFFERINGS:

Many savvy customers, particularly younger customers, don't want to make contact with your business when they have a problem. Instead, they want to self-diagnose the problem using resources they find online. Thus, to maintain a competitive edge over the long term, you want to be continuously expanding and enhancing your self-service offerings—through more knowledge articles, Q&A-style community forums, and dialogue boxes embedded into products and services that pop up with helpful hints and troubleshooting suggestions. As these features become more commonplace, customers will increasingly come to expect them—and expect that the features will be intuitive, helpful, interconnected, and comprehensive.

10.4 ENABLE YOUR CUSTOMERS TO SHARE THEIR EXPERIENCES WITH YOUR COMPANY:

As social media explodes in popularity, you want to not only have a strong presence on social media that allows your customers to find

you, but also to proactively invite all of your customers to share their experiences with your company on social media. You can do this by explicitly building the option to share into your sales funnel, and you can even incentivize sharing with contests and prizes for participants. Social media does not offer companies total control over these messages, but the way that customers engage with your business over social media will become increasingly vital for maintaining relevance and prominence in an increasingly social-connected world. The strategic value for your company is that these social media impressions are likely to be seen by like-minded audiences who will be interested in your products and services.

A company that isn't continuously innovating is slowly dying; innovation is at the heart of the battle for long-term customer loyalty. To maintain your competitive edge and drive new profits, you want to focus on multiple strategies for supporting innovation within your contact center, including improving your ability to proactively initiate contact with your customers, using the Internet of Things to gain new customer insights, boosting your self-service offerings, and enabling your customers to share their experiences with your company in a way that will get others interested in your products and services.

How confident are you in answering these questions?

- How does your contact center team view innovation in relation to CX?
- What would be an effective strategy for proactively initiating contact with more of your customers?
- How could you take advantage of the Internet of Things to collect additional insights on your customers?
- How could you use social media exposure that you can't directly control to your strategic business advantage?

Final Thoughts

ServiceNow CSM has the potential to rapidly and comprehensively transform your customer service operations from the inside out. The *10 Pillars of CSM Success for Customer Service Leaders* has given you a clear roadmap outlining how to focus your thoughts, goals, and priorities for CSM modernization going forward.

Don't let this book sit on your computer desktop, forgotten amid your other competing priorities. Digest and internalize the many specific tips and advice Crossfuze's CSM experts shared with you, so that you can propel your organization forward with speed, agility, and confidence.

As you plot your contact center's modernization journey using ServiceNow CSM, remember that you don't need to go it alone. This book hopefully has helped you appreciate the many moving parts and priorities you must juggle to orchestrate an optimal CSM transformation. Crossfuze, a leading ServiceNow Gold Partner, can help you on your journey to CSM modernization. We would be honored and excited to share with you the many lessons learned and the countless successes we have experienced from our 500+ ServiceNow implementations and 25+ years operating contact centers for enterprise organizations.

Made in United States
Troutdale, OR
11/03/2026

56845124R00043